Lea Weller BA

The Introduction of a Compulsory DNA Database

GRIN Verlag

Bibliografische Information der Deutschen Nationalbibliothek:

Die Deutsche Bibliothek verzeichnet diese Publikation in der Deutschen National-
bibliografie; detaillierte bibliografische Daten sind im Internet über http://dnb.d-
nb.de/ abrufbar.

Imprint:

Copyright © 2008 GRIN Verlag GmbH
Druck und Bindung: Books on Demand GmbH, Norderstedt Germany
ISBN: 978-3-656-54340-4

This book at GRIN:

http://www.grin.com/en/e-book/264567/the-introduction-of-a-compulsory-dna-
database

GRIN - Your knowledge has value

Der GRIN Verlag publiziert seit 1998 wissenschaftliche Arbeiten von Studenten, Hochschullehrern und anderen Akademikern als eBook und gedrucktes Buch. Die Verlagswebsite www.grin.com ist die ideale Plattform zur Veröffentlichung von Hausarbeiten, Abschlussarbeiten, wissenschaftlichen Aufsätzen, Dissertationen und Fachbüchern.

The Introduction of a Compulsory DNA Database

The DNA database currently holds DNA samples of convicted criminals, suspects, witnesses, victims, children and people who were not charged with an offence. An article written in the telegraph states that, once on the database, your DNA profile cannot be removed. (George Jones, 2006). Ben Quinn of the Guardian explained that, a call for a nationwide DNA database has been made. (Ben Quinn, 2008) In this report the DNA database is discussed showing issues, both for and against the database. Certain issues raised are social and ethical issues, medical and economic issues and issues with the crime system.

With a DNA database in place, there could be instant recognition after a crime, if DNA evidence was left at the scene. For the argument against, problems can occur with DNA transfer. In an article in The Sunday Times, Professor Allan Jaimeson explains that If DNA is found at a crime scene that has been unintentionally transferred by another person, this could cause a conviction of an innocent person being wrongly accused for example: person A shakes hands with person B. Person B gets into a car that was later stolen and used in a crime, person A's DNA would be in that car even though they had not been involved in the crime, and may be charged with the offence. (Richard Woodsands and Daniel Foggo, 2008) Evidence could also be planted at a crime scene by someone else e.g. a few strands of hair. If corrupt police officers investigating a crime had access to the DNA samples they could plant evidence at a scene just to make sure they got a conviction. Cold cases could be re-opened and the DNA samples can be analysed again to find the criminal, if they were on the database. (George Jones, 2006).

There is also an issue with illegal immigrants entering the country and committing crime. The database would not hold their DNA profile, so they would not be caught.

The DNA database may reduce the police time, resources in dealing with a case and save on money used carrying out the investigations. Mark Henderson, the science editor of Times Online, states that the cost of setting up the compulsory DNA database is estimated at £700 million. (Mark Henderson, 2007) Where would this money come from? The database could save lives and victims in the future. For example: if a victim was murdered, and the murderer left DNA evidence at the crime scene, the murderer would be caught and named straight away, so they could not murder again. So the system would save lives, although as previously discussed, there is the issue of unintentional DNA transfer or evidence planting.

The DNA database would not be used by anyone unless authorised to do so. Training would be given to those who will need to access the database and interpret the evidence e.g. police, lawyers, jurors, judges, scientists, and in the case of medical research doctors and research scientists.

The police will be given training on how to use the DNA evidence with other evidence found. They would need to conduct an investigation and not just rely on the DNA evidence, as there are worries of the police becoming 'lazy' in their investigations and relying on the DNA evidence found at the scene of the crime. The public would also perceive that the police would catch the criminals of a particular offence straight away, and question "why have you not caught them yet"?

The security would be significantly increased, so only people with authorization and training can access the database. But even with increased security there is always a risk of the security being breached.

How would we enforce the DNA database? Starting a DNA database would require the public coming forward to give a sample of their DNA. Not every person in the UK would willingly give a sample of their DNA. To enforce this there would need to be a new law brought in. An alternative approach would be to take blood from a child as soon as it is born at the hospital or when the birth is registered. The idea of taking DNA samples from a new-born is taking away that Childs right to choose, yet if it is the law, then their right to have this choice has already been taken away. Religious issues and beliefs could stop a person from giving blood. If religious issues were to stop that person from giving blood, then there would be a non-evasive procedure. These procedures could include taking swabs and hair samples. As the parents will be in charge of the Childs medical information, they could choose to have tests done to show any future illness or life-threatening disease. If necessary dietary, treatment and lifestyle conditions could be enforced until the child is old enough to take charge of their medical state. Within the medical issues, a lot of the population would not want to know if they had a serious illness or disease, or were to eventually suffer from one. The susceptibility of disease would most likely be too hard to measure from an infant. So they would need regular testing throughout life.

If you had the knowledge of your medical condition, an insurance company could refuse to give you an insurance plan, due to your condition or future condition. In this case, regulations would be put in place to prevent insurance companies discriminating against you if you have or will have a life threatening illness.

Scientific researchers would need to gain permission from a person whose DNA sample they would like to use in a research project. They may not always get permission so the research may fail due to limited samples and results. Research into many aspects could help find cures or treatments for diseases. A DNA database would help find matches for transplants such as bone marrow, organs, blood. So donors would not be as hard to find as they are now. When DNA samples are taken, you would be given the choice of being a donor or not.

In the future it could be made so you would not be able to access anything without being on the database e.g. medical attention, benefits, getting a job, obtaining a passport and drivers licence etc.

How many years would it take to set up the DNA database? There is no answer for this. No one knows how long it could take.

The introduction of the DNA database was refused on ethical and legal implications. Even with the database being so useful in solving crime. An article in the New Scientist, addresses the fact that people are worried about liberty, personal privacy and autonomy. (New Scientist, 2007). But I believe that if a database was to be introduced the argument for outweighs the argument against. In this respect I am for the compulsory DNA database. It will solve more crime through the database. Even though mistakes will be made it would be no different as today. Today people are still being wrongly accused of crime, even without the DNA database. Medical research could find new treatments and allow us to start treatment for conditions as early as possible. If the whole of the UK is on the DNA database- there is no discrimination. We are all in the same situation.

References

BBC News (2007) All UK 'must be on DNA database' BBC News
http://news.bbc.co.uk/1/hi/uk/6979138.stm (Accessed 19/11/2008)

Henderson. M (2007) DNA database 'puts innocent under suspicion' Times Online
http://www.timesonline.co.uk/tol/news/uk/crime/article2477559.ece (Accessed 19/11/2008)

Jones. G (2006) DNA database 'should include all' Telegraph
http://www.telegraph.co.uk/news/uknews/1532210/DNA-database-'should-include-all'.html
Accessed (19/11/2008)

Lewis. J (2008)Police turn up pressure for compulsory DNA database as Yard 'uses DNA to
nail Stephen Lawrence killers' Daily Mail, http://www.dailymail.co.uk/news/article-
517837/Police-turn-pressure-compulsory-DNA-database-Yard-uses-DNA-nail-Stephen-
Lawrence-killers.html (Accessed 19/11/2008)

New scientist (2007) Universal DNA database would make us all suspects, New Scientist
http://www.newscientist.com/article/mg19526223.300-universal-dna-database-would-make-
us-all-suspects.html Accessed (19/11/2008)

O'Brien. C (2002) DNA database "should include every citizen" New Scientist
http://www.newscientist.com/article/dn2792-dna-database-should-include-every-citizen.html
(Accessed 19/11/2008)

Quinn. B and agencies (2008) Calls for compulsory DNA database rejected, Guardian,
http://www.guardian.co.uk/politics/2008/feb/23/ukcrime?gusrc=rss&feed=networkfront
(Accessed 19/11/2008)

Ramde. D (2008) Wis. Governor Unveils Gene Research Triangle, International Business
Times, http://www.ibtimes.com/articles/20081010/wis-governor-unveils-gene-research-
triangle.htm (Accessed 27/11/08)

The First Post (2007) A compulsory DNA database The First Post
http://www.thefirstpost.co.uk/8497,opinion,should-we-have-a-dna-database (Accessed
19/11/2008)

Woodsand. R and Faggo. D (2008) Should Britain have a compulsory DNA database? The
Sunday Times, (2008) (Accessed 17/11/08)